Carbon Sequestration Pilot Program: Implementation and Next Steps

Progress Report

February 2009

U.S. Department of Transportation
Federal Highway Administration
Office of Natural and Human Environment

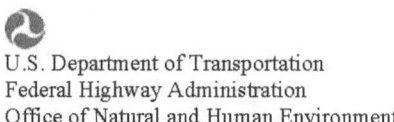
U.S. Department of Transportation
Research and Innovative Technology Administration
John A. Volpe National Transportation Systems Center

Acknowledgments

This report was prepared by the U.S. Department of Transportation Federal Highway Administration (FHWA) and John A. Volpe National Transportation Systems Center (Volpe Center). The project team included Stephen Earsom of FHWA's Office of Natural and Human Environment and Carson Poe of the Volpe Center's Planning and Policy Analysis Division.

The project team wishes to thank the numerous state Departments of Transportation (DOTs) staff members—especially those at New Mexico DOT—who eagerly provided their time and expertise, better enabling FHWA to explore the possibility of a program to sequester carbon along highway rights-of-way.

Contents

ACRONYMS AND DEFINITIONS 1

EXECUTIVE SUMMARY 1

INTRODUCTION 2
    Report Purpose 2
    Background 2

PILOT PROGRAM IMPLEMENTATION AT NEW MEXICO DOT 4
    NMDOT's Participation 4
    Key Stakeholders and Their Roles 4
    Carbon Sequestration Process Development 6
    Lessons Learned and Suggestions 11

MOVING FORWARD 14

APPENDIX A.   SELECTION CRITERIA AND TOP STATE DOT RANKINGS 16
    Essential Selection Criteria 16
    Qualitative Criteria 16
    Considerations for State DOTs 17

Acronyms and Definitions

ARS	Agricultural Research Service
CCX	Chicago Climate Exchange
CSPP	Carbon Sequestration Pilot Program
DOT	Department of Transportation
FHWA	Federal Highway Administration
GHG	Greenhouse gas
HEPN	Office of Natural and Human Environment
MOU	Memorandum of Understanding
NEE	Net Ecosystem Exchange
NMDOT	New Mexico Department of Transportation
NMSU	New Mexico State University
NRCS	Natural Resources Conservation Service
RGGI	Regional Greenhouse Gas Initiative
ROW	Right-of-way
SAFETEA-LU	Safe, Accountable, Flexible, Efficient Transportation Equity Act: A Legacy for Users
USDA	U.S. Department of Agriculture
Volpe Center	U.S. Department of Transportation John A. Volpe National Transportation Systems Center
WCI	Western Climate Initiative

Executive Summary

In July 2008, the FHWA's Office of Natural and Human Environment (HEPN) selected the New Mexico Department of Transportation (NMDOT) to participate in a carbon sequestration pilot program (CSPP). Through the CSPP, FHWA intends explore the feasibility of state DOTs reducing and sequestering greenhouse gas (GHG) emissions in vegetation within highway rights-of-way (ROW).

Under the pilot program, NMDOT is undertaking a four-year, $2 million research project to quantify the amount of atmospheric carbon that grasslands along highway ROW can sequester. The protocol that will result should be applicable to DOTs nationwide, an important consideration since part of the pilot program's success centers on the DOT's ability to measure then divest the carbon captured. Options for divestiture are (1) selling carbon credits on an appropriate GHG market or registry for revenue, (2) using carbon credits to offset the DOT's emissions, or (3) using the credits toward meeting statewide objectives.

NMDOT's initial experience with this ground-breaking effort has revealed a number of lessons useful to other DOTs and FHWA Division Offices in evaluating the viability of carbon sequestration practices in lands they control. Some key findings are:

- Information thought to exist on carbon sequestration protocols for grasslands that are not grazed are not available upon closer inspection;
- The number of verifiers currently in the western U.S. is limited;
- Identify, manage, and mitigate risks that could affect the ability to trade carbon credits;
- Determine how to supplement the maintenance and operations records that are currently kept;
- Agencies' district offices sometimes act on their own authorities, resulting in dissimilar decisions within a state regarding similar topics;
- Reducing emissions through modified management practices could contribute more to meeting GHG goals than carbon sequestration.
- Implement ecologically-sound land management practices holistically (e.g., planting climate- and season-appropriate seed species statewide), as sequestering carbon in one region does not permit deficient land management practices in another;
- There are questions as to what rights a DOT has within the easements it has with federal partners. Accordingly, FHWA must clarify ownership of carbon credits on lands that DOTs do not own in fee simple.
- Improved vegetation management can result in benefits independent of carbon sequestration. Examples include reduced soil erosion due to increased vegetation, and lessened fuel consumption and emissions releases resulting from reduced mowing frequency and intensity.

This progress report expands on these and other lessons NMDOT has learned to date. NMDOT's insights are expected to help accomplish a similar program at a broader scale and/or assist in making well-informed decisions related to carbon sequestration in highway ROW during reauthorization of the next transportation bill. Consequently, FHWA is currently soliciting a second DOT from a more humid state to complement NMDOT's participation.

Introduction

Report Purpose

This report's purpose is two-fold. First, it documents the elements likely required to implement a carbon sequestration process tailored to state Departments of Transportation (DOTs). Findings are based on the challenges, key lessons, and preliminary results from New Mexico DOT's initial five months of implementing the Federal Highway Administration's (FHWA) Carbon Sequestration Pilot Program (CSPP). Although communication of the progress and outcomes should continue through the life-cycle of the pilot program, it is expected that the body of knowledge formed from NMDOT's early experience can help other state DOTs evaluate the viability of carbon sequestration practices in lands they control.

The CSPP is also considered a demonstration project. As such, this report offers preliminary recommendations and considerations that can assist FHWA and other federal agencies in making well-informed decisions related to carbon sequestration in highway ROW during discussions surrounding the reauthorization of the nation's next transportation bill. Since the NMDOT effort is ground-breaking nationally, and perhaps globally, it is expected that NMDOT's experience can provide insights useful in determining whether efforts to accomplish a similar program at a broader scale are practical.

Background

In recent years, the United States Department of Transportation (USDOT) has sought to better understand global climate change and the transportation sector's impact on the issue. The current transportation bill, SAFETEA-LU, has provided an opportunity for state and federal agencies to conduct research on innovative practices that may reduce transportation-related green-house gas (GHG) emissions and by extension, the transportation sector's impact on climate change. In 2008, FHWA's Office of Natural and Human Environment (HEPN) and Office of Project Development and Environmental Review (HEPE) created the CSPP in an effort to define a process or program that could demonstrate the value of sequestering or capturing carbon from the highway right-of-way (ROW) through modified maintenance and management practices. The CSPP was designed to help a state DOT reduce emissions and maintenance costs, generate revenue on an appropriate market, and foster ancillary environmental services benefits, such as reduced erosion, better retention of stormwater in soil, enhanced ability to hold snow, and improved wildlife habitat.

Biological carbon sequestration is one strategy for using vegetation to reduce carbon dioxide (CO_2) in the air. Generated from combustion of fossil fuel, CO_2 comprises roughly 80 percent of human-made GHG contributing to global climate change. Deforestation and land degradation have reduced the ecosystem's ability to remove CO_2 from the atmosphere. "Carbon sequestration" is the capture and storage of CO_2 from the atmosphere. Along highways, vegetation growing near the roads can absorb CO_2. In plain terms, plants take in CO_2 during photosynthesis and store, or sequester, it in their leaves and stems or transfer it to soil via their roots. Though plants release CO_2 back to the atmosphere when they die, long-term land management can result in the sequestration of predictable and significant carbon volumes.

In addition to ecological benefits, carbon sequestration can have meaningful economic impacts. Markets for trading "carbon credits," or offsets, are in the early stages of development.[1] The Chicago Climate Exchange (CCX), launched in 2003, is one such market. It offers a "legally binding integrated trading system" to reduce GHG emissions by facilitating the sale of surplus carbon allowances or the purchase of emissions contracts by its members. Entities seeking to voluntarily reduce their emissions can buy and sell the allowances to offset their excess emissions (see Figure 1 for historical carbon prices). Under a national "cap and trade" system, which is an emissions reduction tactic using a national emissions ceiling that is reduced over time, participation in a carbon market would not be voluntary for those with emissions greater than the established threshold.

Figure 1. CCX Carbon Financial Instrument Contracts Daily Report*
Source: Chicago Climate Exchange

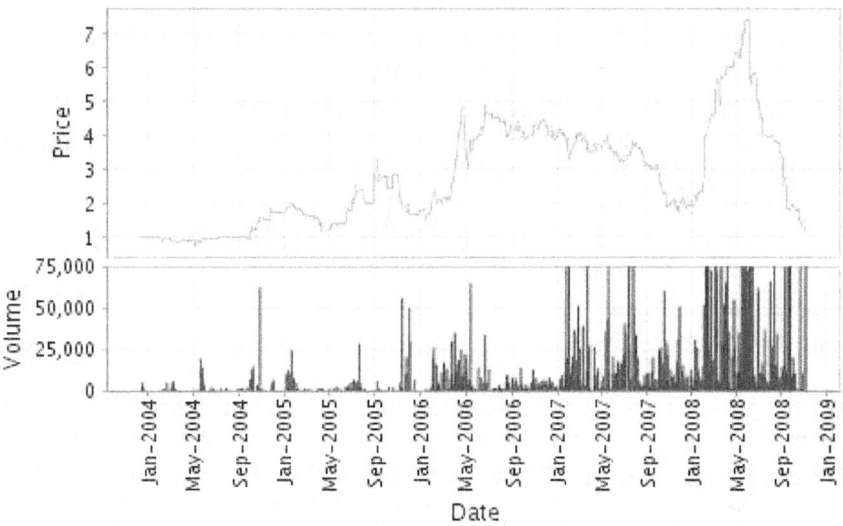

*Prices are reported in dollars per metric ton of CO_2.

Related projects, such as the Regional Greenhouse Gas Initiative (RGGI) and Western Climate Initiative (WCI), target specific sectors for emissions reduction. RGGI is a coalition of 10 northeastern and Mid-Atlantic States that uses a market-based cap-and-trade approach to reduce CO_2 emissions from the power sector. WCI is a collaboration of seven U.S. governors and four Canadian Premiers. Much like RGGI, the WCI identifies and employs cooperative ways to reduce greenhouse gases in the region, focusing on a market-based cap-and-trade system. Both RGGI and WCI have established coalitions of states or jurisdictions to engage in coordinated market-based emissions reduction.

---

[1] A carbon credit, the universal metric used in GHG accounting, is defined as one metric ton of CO_2 or its equivalent.

Pilot Program Implementation at New Mexico DOT

NMDOT's Participation

The process for selecting a participant for the CSPP was iterative, requiring several cycles of "narrowing the candidate field." Initially, FHWA and the Volpe Center scored all 50 states and Puerto Rico on the basis of available ROW acreage, the state's potential to legally participate in carbon emissions trading, and various additional criteria (see Appendix A). The top-scoring state DOTs were notified by their states' respective FHWA Division Office of the availability of the pilot program and were invited to participate in a second round of the selection process. A questionnaire intended to acquire detailed estimates of ROW acreage was then sent to six candidate states. Responses were collected from each via telephone interview.

After the interviews, NMDOT was selected as one of two "finalists." Staff from FHWA and the U.S. DOT Volpe National Transportation Systems Center (Volpe Center) visited the finalists in May and June 2008 to gather supplementary data for selecting a CSPP participant. FHWA chose NMDOT to participate based on three primary factors:

- Its alignment with the selection criteria; for example, NM has many miles of highway, fairly wide ROWs, and a broad range of native grasses, shrubs, and trees adapted to arid conditions;
- Information gathered during the site visit; and,
- Its level of interest in being involved in the pilot program.

Since its selection in July 2008, NMDOT's CSPP interest and dexterity in developing the institutional framework necessary to support carbon sequestration activities have been consistent. In order to learn more about the DOT's efforts to advance the CSPP, FHWA and Volpe Center returned to NMDOT from October 6–9, 2008. During the site visit, NMDOT convened several meetings with those who have been and are likely to be important contributors in the enterprise to implement the CSPP. The following sections summarize the information gathered during those discussions.

Key Stakeholders and Their Roles

NMDOT's CSPP team consists of staff from the Department's Environmental Design Division. Their CSPP participation required 550 total labor hours between July 2008 and December 2008, including the input of NMDOT executives. A portion of this time has been used to learn about development of a protocol for quantifying and verifying carbon sequestered in grasslands that are not grazed (since no protocol for this type of land currently exists). Estimates of labor requirements for future carbon sequestration efforts would not need to factor in time to develop existing protocols, as they are reusable.

From the earliest stages, it was apparent to the NMDOT project team that input from a variety of skills and expertise would be necessary to implement carbon sequestration practices. Key internal and external stakeholders that have been identified to date include:

Internal

- *NMDOT leadership*
- *Environmental staff*
- *Maintenance staff*
- *Operations staff*
- *Geographic information systems (GIS) staff*
- *Planning Staff*[2]

External

- *Carbon aggregator.* An organization can gain access to the CCX through a "carbon aggregator." These entities are brokers who aggregate acres that can be efficiently traded in large trading blocks. In other words, an aggregator serves as the administrative representative for multiple offset-generating projects on behalf of multiple project owners who individually have relatively few credits to sell. Aggregators are responsible for enrolling and certifying land in the trading program and ensuring that the enrolled acres conform to market standards. They charge a fee for their services that can be as much as 50 percent of the credit value.

    NMDOT plans to explore opportunities to become its own aggregator through partnering with another state(s).[3] A state DOT might consider doing so if it were uncertain as to whether enough CO_2 tonnage could be sequestered on its own for it to make sense economically. Instead of hiring a third party aggregator, a partnership with other states could help ensure the volume of carbon sequestered was significant enough to generate meaningful returns on a market (i.e., by avoiding the costs and commissions of the third party entity).

- *Carbon verifier.* Carbon aggregators work in concert with "carbon verifiers," also critical to entering a carbon credit market, to ensure that enrolled land has followed the established protocol in claiming carbon credits. Verifiers charge a service fee that is deducted from the annual sale proceeds to cover expenses associated with managing the program.

- *Personnel with working knowledge of carbon dynamics.* Carbon dynamics describe the process by which carbon moves through an ecosystem. Since carbon trading platforms, such as CCX, only issue credits for carbon that has been sequestered above and beyond baseline sequestration levels, it is critical to understand how much carbon is being sequestered with no change in land management practice. NMDOT is considering opportunities for partnering with the U.S. Department of Agriculture (USDA) Agricultural Research Service's (ARS) Jornada Experimental Range to develop a systematic approach for establishing a carbon baseline.

    NMDOT is working with the Jornada Experimental Range to develop a protocol for bringing carbon credits that were generated from grasslands not grazed to market. This is necessary only because no protocol has been established for this type of land. Once the protocol is developed, it could be used for highway ROW nationally. Protocols for crediting carbon reductions on other lands/resources, such as rangelands that are managed/grazed, exist and do not need be recreated.

---

[2] Staff members from NMDOT's Statewide Planning Section serve on the Governor's Climate Change Advisory Group. Staff from the Environmental Design Division working on the CSPP may become involved in various technical aspects of the committee's work.

[3] As an alternative, should NMDOT decide to sell carbon offsets it could register with CCX, for example, as an "offset provider." Members designated as offset providers register and sell a project's offsets directly on the CCX.

- *Seed provider.* NMDOT currently obtains seeds for post-construction reseeding from private seed producers with whom NMDOT has purchase agreements. In some cases, the construction contractor for a project will purchase the seeds. The Natural Resources Conservation Service (NRCS) Los Lunas Plant Materials Center develops the seed that commercial growers use for production. NMDOT is working with the NRCS to evaluate and update the department's revegetation specifications to help ensure the season- and ecoregion-appropriate seeds are planted at the proper times after construction.

- *Federal land management agencies.* NMDOT has identified a need to work with the federal land management agencies to understand the DOT's rights within easements it has with federal partners. This is important because not all federal lands' ROWs are held in fee simple. It is currently unknown how ownership rights to lands a DOT manages may affect the process for selling carbon credits generated on those lands. Discussions at the federal level regarding this and other climate change-related topics are on-going.

- *WCI.* In 2007, New Mexico joined with Arizona, California, Montana, Oregon, Utah, and Washington, as well as the Premiers of the Canadian provinces of British Columbia and Manitoba to create the WCI. These members have established a regional GHG reduction goal and are now developing the design for a cap-and-trade program to help achieve that goal. NMDOT plans to work with the WCI to understand how a WCI cap-and-trade program could affect NMDOT's options for selling carbon credits. There may also be opportunity for NMDOT to participate on a WCI sub-committee.

- *New Mexico Environment Department.* The Environment Department is the lead state agency in the Governor's Climate Initiative programs. The department is responsible for monitoring each agency's compliance activities and is New Mexico's signatory and representative to the CCX.

- *New Mexico Climate Change Advisory Group.* Established in June 2005 by the Governor, the Climate Change Action Council is a diverse group of stakeholders from across New Mexico that reviews and provides recommendations to the Governor's office regarding climate change policy.

NMDOT's CSPP team has worked to engage appropriate stakeholders and experts. The response has been positive and the audience receptive. However, aside from USDA-ARS' Jornada Experimental Range staff, most agencies have indicated that they are relatively unacquainted with the specifics of biological carbon sequestration and have been content to request that NMDOT keep them informed as NMDOT proceeds. This is important because it indicates that NMDOT is shaping a new process. To date, no external policy obstacles have been identified, and others are eager to learn from the NMDOT experience.

Carbon Sequestration Process Development

When FHWA announced its intent to support a pilot program for carbon sequestration along highway ROW, the assumptions as to the process a state DOT might employ were unfixed. It was expected, however, that a DOT would (1) quantify the acreage available for carbon sequestration, (2) estimate the vegetation costs for altered planting practices, (3) estimate the carbon credits available from the enhanced management techniques, and then (4) identify a verifier that would confirm the amounts of carbon sequestered, enabling participation in an appropriate trading market. While still in the early stages of CSPP implementation, NMDOT has found these to be guiding steps, influenced by a variety of factors. This section documents NMDOT's experience to date to develop a process to quantify, verify, and market carbon sequestered, including challenges and concerns encountered.

1. *Quantification of Land Available*

During the state selection stage of the CSPP, HEPN surveyed candidate states to gather early approximations of unpaved ROW amounts that the DOTs controlled. Once NMDOT was chosen to participate, the Department began working with the Division Office, HEPN, and other partners to refine the initial estimate in order that appropriate lands for inclusion in the sequestration project can be identified. A lesson NMDOT learned early on is that to participate in the CCX, land should be measured in hectares, not acres.

The approach that NMDOT has taken to quantifying hectares available has involved collecting detailed roadway data from its Transportation and Highway Operations office, including miles of highway controlled, widths of unpaved ROW, and current management practices on those ROWs. This information was difficult to obtain in an expeditious manner. Data resided in ROW maps in project files and not in digital vector format that would facilitate statistical manipulation and seamless integration into GIS. The translation of the ROW information into a GIS medium is an on-going challenge, which likely requires considerable effort in a time when many funding resources are limited.

Nevertheless, the roadway information collected will be directly tied to estimations of carbon potentially sequestered, and thus the lands' revenue-generating potential. Should a DOT decide to market carbon credits, the collective ROW available for carbon sequestration would need to be able to produce enough carbon credits to generate sufficient revenue to more than offset costs. In other words, after considering the current market price of a ton of carbon, sufficient amount of unpaved ROW where land management practice modification and/or enhancement is possible should be available to make sequestration profitable. If the carbon sequestered was not used for credits, another approach would be to use the sequestered carbon internally or for other state agencies as offsets to meet agency/state GHG emissions reduction targets.

NMDOT plans to collaborate with internal offices, such as those responsible for safety and vegetation management, to better understand the variety of considerations that may affect the feasibility of where changes in land management practices could occur. GIS specialists will likely be helpful in analyzing existing maps, aerial photography, and other geospatial data related to the Department's ROW.

2. *Estimation of Vegetation Costs*

Roadside ROW accounts for more than 10 million acres of land in the U.S.[4] This land requires care to meet a variety of vegetation management objectives, such as:

- Maintenance of a safe ROW by providing clear sight distances;
- Assurance of water quality and provision of adequate drainage in roadway ditches;
- Erosion control improvement;
- Reduction of fire hazard and provision snow drift control;
- Wildlife habitat improvement;
- Mowing and spraying reductions;
- Enhancement of natural beauty;
- Noxious weed control; and,
- Protection of natural heritage.

---

[4] *The Nature of Roadsides: And the Tools to Work with It.* FHWA. Publication No.: FHWA-EP-03-005 www.invasivespeciesinfo.gov/docs/plants/roadsides/index.htm

Prior to implementing its Integrated Vegetation Management Program, which was established approximately 10 years ago and combines preventive, mechanical, and chemical vegetation management techniques, NMDOT managed vegetation primarily through its mowing operations. With the establishment of the CSPP, NMDOT is investigating ways to improve and supplement the management program in order that any carbon offsets generated could clearly demonstrate "additionality." The concept of additionality answers the question of whether the project is reducing emissions regardless of the prospect of offset revenues. Demonstration of additionality is required to enter a carbon trading market. Given these factors, NMDOT is considering the following enhancements to its vegetation management practice:

Seeding and Planting Practice
At present, NMDOT reseeds an area disturbed by construction once the construction is completed. One primary reason for this is that Clean Water Act Section 402 General Stormwater Permit conditions are not fulfilled until 70 percent re-growth is obtained. However, in ecological terms, construction projects and seeding, depending on the time of construction, may both be better served as two separate projects. For this reason, NMDOT is looking at how it might separate revegetation efforts from construction and conduct seeding operations at times of the year that will maximize growth and survivorship by taking advantage of precipitation patterns.

In addition to the timing of plantings, the composition of seed species that NMDOT plants could be modified. The DOT has price agreements for common and frequently conducted maintenance activities such as guardrail repair, signing, striping, and pavement preservation. The NMDOT, which utilizes a variety of seed mixes based on different vegetation/habitat types around the state, is now exploring options to update its revegetation specifications and to determine the cost of augmenting the existing vegetation within the ROW with new plant materials. Dr. Joel Brown at NMSU believes that revised reseeding practices could significantly increase carbon sequestration rates. Some have estimated that by establishing perennial vegetation, soil carbon levels can reach 95 percent of levels achievable by undisturbed land. Additionally, a seed mix that includes legumes, which annually fix nitrogen, can provide a natural fertilizer that will result in 20–30 percent increases in total carbon sequestered.

If NMDOT is successful in creating a separate revegetation program, it would be incorporated into the stormwater pollution prevention plans (SWPPPs), which contractors generally develop on a project-by-project basis.

Stormwater Management
Generally, current stormwater management practice at NMDOT involves diverting stormwater into existing streams. Depending on policy decisions, it is likely possible to change the configuration of ditches to channel stormwater sheet-flow to the unpaved ROW, providing water (potentially storable in some locations) for vegetation located there.

Mowing
Given that highways can generate significant amounts of noxious weeds (e.g. inadvertent releases from trucks moving hay across states/eco-regions), there is a very concerted effort in New Mexico to control and prevent noxious weeds. For example, one NMDOT District and approximately 10 other agencies have signed a Memorandum of Understanding (MOU) that describes how the organizations can work together to address noxious weeds issues. In the absence of funding to start a major anti-weeds program, common vegetation management practice at NMDOT currently focuses on spraying herbicides in areas where mowers cannot reach. Herbicide application in New Mexico costs roughly $20–100 per acre. Yearly spraying begins in March of each year and is performed in selected areas through July.

Traditionally, mowing, where practicable, has occurred from fence line to fence line on NMDOT's ROW. In early 2008, NMDOT began a practice of single pass mowing at 8-foot width in wide medians. This approach, which has created considerable reductions in fuel costs[5] and emissions, could be a significant source of additionality when verifying carbon sequestered (see section below for more). The Department's records for the number of highway miles it has reduced mowing on could also be useful in quantifying ROW available for sequestration.

3. *Estimation of Carbon Credits*

Assuming that a baseline is established, the number of carbon credits to be sold would need to be estimated and a protocol developed for doing so. To date, NMDOT has not begun this process. This is primarily due to a policy decision to not address carbon sequestration from woody vegetation (for which protocols already exist). Given potential safety concerns of planting trees along the roadside, NMDOT has decided to explore carbon sequestration in the grasslands along its ROW. Because the ROW grasslands are not grazed and no protocol currently exists for grasslands that are not grazed, NMDOT is undertaking a 4-yr, $2 million research project to determine sequestration rates along highway ROW. If results of this effort indicate economically-viable amounts of carbon will be sequestered, NMDOT will develop a quantification protocol and submit it for approval to CCX or another carbon trading platform. This would be of significant use, since the protocol would be broad and applicable to DOTs across the nation (i.e., the protocol, which could be used for highway ROW nationally, would not need to be recreated).

NMDOT also plans on estimating the emissions reductions possible from reduced mowing and other improved management practices. In some cases, reducing emissions through modified management practices can contribute to meeting GHG goals at least as much as carbon sequestration.

Preliminary lessons in this area are:

- It is important to understand the difference between "tons of carbon" and "tons of carbon dioxide equivalents (CO_2e)." Most markets trade in the latter. Carbon dioxide equivalents offer a standard measurement used to compare the emissions from various GHG based upon their global warming potential. One ton of carbon equals 3.3 tons of CO_2e.

- One Jornada Experimental Range study estimated the average annual net ecosystem exchange (NEE) of carbon for grasslands in Las Cruces, NM. In four out of six years, NEE equaled 1, indicating that carbon had been moved into the soil.

- The potential for tree biomass in U.S. forests to sequester carbon is approximately 1–10 tons of C/ha/year. An informal estimate based on previous research is that tree biomass in highway ROWs could sequester on the order of 0–5 T of C/ha/year.[6]

Presently, NMDOT needs more data to make more definitive estimates. Only order-of-magnitude estimates would be possible at this time.

---

[5] From 2005 to 2007, NMDOT purchased 3.0 million gallons of gas and diesel. In 2008, this volume will be down to 2.6 million gallons. Along these lines, NMDOT currently has a pilot program to determine how biodiesel use affects its fleet and equipment. If biodiesel is shown to be effective, NMDOT anticipates moving completely to biodiesel, further reducing fossil fuel use and emissions.

[6] Estimates are from presentation by Dr. Joel Brown, Jornada Experimental Range, at meeting with NMDOT in October 2008.

4.   Identify, Selecting, and Working with a Verifier

NMDOT is engaged in a process to understand as much as possible about carbon verifiers, their requirements, and how it will meet those requirements. This process is crucial to bridging the gap between the ecological and biological analyses and the economic considerations made. As a first step, NMDOT conducted an Internet search of potential verifiers.

Initially, it was challenging for NMDOT to identify candidates. The number of verifiers currently in the western U.S. is limited, and few have specific expertise in New Mexico's ecoregions, specifically those where native grasses grow.[7] A lack of local/regional scientific expertise from the first potential verifier NMDOT contacted required the agency to conduct a more extensive search for a more suitable partner. A companion issue arose with NMDOT's initial lead: verifiers will likely want assurance that enough carbon can be sequestered for it to be economically feasible for the verifier to participate. Since verifiers receive a commission of carbon credit sales, usually 5–20 percent, project sites must sequester a volume of carbon large enough to produce a meaningful commission for the verifier. To date, no verifiers have been contacted because none have been certified for the type of ROW offset potentially generated. NMDOT has researched the CCX's verifier certification process and will include one adapted to ROW offsets as protocols for the CCX and WCI are developed.

Carbon offsets must also clearly demonstrate "additionality" and have a realistically calculated baseline and emissions reduction projection. As NMDOT held discussions with the potential aggregator, it became apparent that the DOT had little to no baseline data related to the Department's emissions and/or soil carbon sequestered in grasslands along the ROW. However, NMDOT may have the GIS data and/or aerial photography that could be used to estimate woody vegetation in a baseline year. Since NMDOT currently does not intend to use woody vegetation to sequester carbon, it has shifted its focus to determining how to develop a baseline for grasslands from which additionality can be measured.

Additional Internet research led NMDOT to the USDA-ARS Jornada Experimental Range located at New Mexico State University (NMSU) in Las Cruces, NM, where, a scientist with prior experience developing carbon verification protocols for the CCX was based. Initial discussions with Jornada Experimental Range experts revealed the importance of including staff that are well-versed in carbon dynamics in any effort to quantify, verify, and sell carbon credits. Absent this knowledge, it would be difficult for participants to accurately assess risks, rewards, and next steps.

Since forging a relationship with the Jornada Experimental Range, NMDOT has also learned more about the process by which its carbon sequestration potential can actually be verified. That process will hinge on a systematic approach that establishes a baseline level of carbon currently being sequestered in the soils, grasses, and woody shrubs of the ROW. Other physical characteristics of the ROW, such as precipitation, soil moisture, and standing crop,[8] can help predict how much carbon can be sequestered. The Jornada Experimental Range has noted that for the CSPP to be feasible, the verification process must show that it does not cost more to store the carbon than that carbon will be worth on the market.

Still, there is uncertainty in how to cost-effectively determine a carbon sequestration baseline. NMDOT may pursue an analysis of aerial images to create estimates, which would later be refined through more detailed multispectral imagery analyses. Another approach being considered is an integrated series of field plot tests, statistical sampling, data gathering, and computer modeling. The field tests would be centered around core sites where data are collected intensively and used to parameterize models. While

---

[7] For a complete list of the CCX's approved offset project verifiers see www.chicagoclimatex.com/content.jsf?id=102.
[8] "Standing crop" refers to the amount of organisms per unit area at a given time.

likely more precise than using imagery, one field test challenge is the high level of variability in the distribution of organic matter in fields. For 95 percent confidence, a field test would require approximately 70 samples per hectare. At a minimum, the costs for each sample (which includes travel to the site, core extraction and preparation, and chemical and statistical analyses) would be approximately $15 per sample. These costs must be balanced against the goal of attempting to detect very small changes – often less than 1 percent – in carbon sequestered each year. The practical challenge is to maintain consistency of samples across a state (e.g. taking the samples as temporally close together as possible).

It is important to note that this research is necessary only because there currently is no quantification protocol for sequestering carbon in grasslands not grazed. Protocols already exist for woody vegetation, and thus, methods for determining baselines, such as aerial imagery analysis, are likely sufficient.

If the measurement challenges can be surmounted, NMDOT can be positioned to better understand the carbon sequestration risks through evaluating expected crediting rate variations in the state's regions. Some risks are:

- The variability of carbon prices
- What federal emissions reductions targets may be, if mandated
- A DOT's outlook on and the state's ecological ability in growing trees in the ROW (trees can double sequestration rates)

Because CCX contracts can range from 5 to 20 years, NMDOT would likely use the baseline data and verifier's analysis to develop an appropriate risk management plan to protect its carbon assets. Although a method for determining a baseline in NM has not been chosen, NMDOT will likely use FHWA recently approved research grant funding to support the effort.[9]

Lessons Learned and Suggestions

NMDOT has discovered several lessons in its efforts to implement the CSPP. These results are intended to inform the decisions of state DOTs, FHWA Division Offices, and their partners on similar future projects. Each of these suggestions, which target state DOTs pursuing a carbon sequestration effort, fall into three broad categories: Staff Lessons, Process Lessons, and Technical Requirements Lessons. In cases where there was overlap among lessons, the suggestions are combined and are presented together.

Staff Lessons

- **Identify and involve those with knowledge of the eco-regions and habitats of the state.** With extensive variability among ecosystems across the country, and in some cases within a state, it is important that personnel with significant understanding of those ecosystems be involved in any carbon sequestration program a DOT may implement. These staff members will likely be responsible for estimating the baseline carbon volume and annual sequestration rates, which are based on soil composition and native plant species distribution in the state. Sound scientific data are needed before a protocol for grasslands that are not grazed can be developed.

---

[9] The NMDOT has teamed with the NMDOT Research Bureau and has received a multi-year grant from the New Mexico Division of the FHWA. The primary goals of this program are to (1) establish the carbon baseline, (2) establish management practices to attain a measurable net increase in carbon sequestration through active management of the highway ROW, and (3) develop applicable protocols for carbon cap and trade systems.

Additionally, NMDOT has found that its counterparts are relatively unfamiliar with the concept of carbon sequestration. While it may not be central to a state DOT's mission, outreach to provide relevant stakeholders a working understanding of the topic may be necessary to implement carbon sequestration activities.

- **Identify and involve those knowledgeable in carbon dynamics.** Expertise in carbon dynamics is most likely found in universities, consulting firms, and/or a state's Department of Natural Resources. In engaging carbon dynamics experts, it is important that state DOT staff members are aware of how the DOT's actions can affect sequestration rates, how protocols are applied and carbon accounting carried out, and what the outlets for selling carbon credits are (e.g. CCX, RGGI, WCI, etc).

It is also important to understand the difference between "tons of carbon" and "tons of CO_2e." Most markets trade in tons of CO_2e.

- **Acquire and maintain support from leadership within the DOT and Division Office.** Carbon sequestration in highway ROWs is not currently a burgeoning discipline at state DOTs. Therefore, an effort to create a program at the margin to do so will require the support of upper management decision-makers (who may also need to be briefed about carbon sequestration). Early during the CSPP's candidate selection phase, NMDOT informed its leadership of the opportunity and has worked to keep it notified of progress and obstacles throughout the process.

Process Lessons

- **Information on carbon sequestration protocols for grasslands that are not grazed is not available upon closer inspection.**

- **Identify, manage, and mitigate risks that could affect the ability to trade carbon credits.** NMDOT has had to determine what uncertainty it is willing to tolerate in pursuing carbon sequestration in its ROW. One uncertainty NMDOT staff has contemplated is drought and in response, development of a drought response plan is being considered. While NMDOT may be creating a protocol (for grasslands not grazed) and process for DOTs to participate in marketing carbon credits, other DOTs will need to assess their own unique risks and develop contingencies for the distinctive risks it faces.

- **Study the maintenance and operations records that are currently kept and determine how to supplement them.** As part of the CSPP, NMDOT is developing an understanding of the types of data (and acceptable surrogates) needed to support a DOT's effort to bring carbon credits to market. Examples are land ownership records, fuel purchases, electricity usage, mowing statistics, etc. Land ownership records are important in determining the number of acres under the DOT's control and thus potentially available for carbon sequestration. Data such as fuel purchases and electricity usage are important because evidence of emissions reductions can be used to meet GHG goals outlined in a contract with a carbon market.

NMDOT has noted that if data required for participation were currently incomplete or not collected, it is likely easy to begin filling these gaps.

- **Agencies' district offices sometimes act on their own authorities, resulting in dissimilar decisions regarding similar topics.** Standard business practices do not always translate from one district to another within an agency. This is important because agreements made with one field office that has jurisdiction in a particular DOT district or region may need to be duplicated in regions under jurisdictions of other field offices.

- **In some cases, reducing emissions through modified management practices, such as reduced mowing, can contribute more to meeting GHG goals than carbon sequestration.**

Technical Requirements Lessons

- **Implement sound land management practices statewide.** Sequestering carbon in one region does not allow for deficient land management practices in another.

- **Convert measurements of ROW acreage available for carbon sequestration to hectares.** To participate in the CCX, land should be measured and reported in hectares, not acres.

- **Plant a climate- and season-appropriate composition of seed species.** In ecological terms, construction projects and revegetation practices, depending on the time of construction, may both be better served as two separate projects. Seeds should be planted when they are likely to receive sufficient precipitation and sunlight to grow without irrigation. In some cases, improved reseeding practices are expected to significantly increase carbon sequestration rates.

- **There are questions as to what rights a DOT has within the easements it has with federal partners.** Coordination at the federal level is likely necessary to establish policy recommendations for non-traditional land management practices, such as carbon sequestration, that state DOTs may undertake on easements it manages. Currently, it is unknown whether carbon credits resulting from a DOT's management practices that are generated on federal lands are possible, and if so, how they can be traded.

Moving Forward

Since January 2007, the U.S. House of Representatives Committee on Energy and Commerce has been one of several groups[10] examining draft climate change legislation. As the committee notes, "politically, scientifically, legally, and morally, the question has been settled: regulation of greenhouse gases in the United States is coming...The only remaining question is what form that regulation will take."[11] The CSPP is a forward-looking effort aimed at informing state DOTs and FHWA Division Offices about one option that may become required in the future.

However, sufficient time has not passed to evaluate NMDOT's ability to design and implement a project based on its capacity to significantly reduce carbon and the associated revenue potential. What is acknowledged is that carbon sequestration will likely not ever be the highest priority for a state DOT. Early results from the CSPP have illustrated several reasons for improved vegetation management that are independent of carbon sequestration. However, if NMDOT demonstrates that improved vegetation management can potentially pay for itself through revenue generation from selling carbon credits, sequestering carbon in highway ROWs could prove to be an attractive complement to existing practice. Similarly, with federal and/or state climate change legislation and regulation likely, carbon sequestration in highway ROWs could be an effective method of assisting in meeting GHG emission reduction targets. In that instance, generating revenue may become secondary to environmental compliance.

With the reauthorization of the transportation bill approaching, FHWA is using the CSPP to assess whether a larger-scale roadside carbon sequestration effort is appropriate when balanced against the uncertainties. Carbon sequestration would be one facet of a holistic and ecologically-sound approach to ROW management practices that are already being carried out.

FHWA is currently soliciting another state DOT to participate in a pilot to sequester carbon in its ROWs, and is hoping to work with a state with ample precipitation and woody vegetation. In this case, quantification protocols, and presumably verifiers, would exist and the DOT would be able to begin taking the necessary steps to estimate its carbon baseline from the outset. The envisioned effort would allow a state DOT(s) and an FHWA Division Office(s) to accomplish the following key three aspects of marketing carbon credits:

(1) *Identification of and coordination with key stakeholders at both federal and state levels*

(2) *Estimation of baseline* – Quantification of ROW available for carbon sequestration and what the physical characteristics and current management practices on the available lands are;

(3) *Estimation of changes to baseline* – Collection of evidence that management changes are linked to changes in carbon sequestered. These measurements could be based on a credible modeling and remote sensing techniques developed as a result of the effort. NMDOT's upfront work to create a model process for roadway carbon sequestration would be transferable. There would be

---

[10] S. 2191, S. 280, and HR. 620 are among the related bills introduced in the 110th Congress.
[11] Committee on Energy and Commerce Memorandum, October 7, 2008.
http://energycommerce.house.gov/Climate_Change/Memo-Climate-Change-100708.pdf

little need to repeat this work. Estimated carbon sequestration baselines, however, would not be transferable. The additional pilot state DOT(s) would need to establish its own baseline.

Appendix A. Selection Criteria and Top State DOT Rankings

Essential Selection Criteria
The following three criteria will be used to evaluate potential state DOT participants.

1. The state DOT selected for the CSPP should own adequate acreage, inside or outside the ROW, for large-scale vegetation planting, maintenance, and evaluation.

One objective of the project is to realize substantial financial incentives for state DOTs above and beyond their initial investment. At $10/ton of CO_2, a plot of 10,000 acres could gross as much as $500,000 annually in carbon credits sold (before registry fees). Under this criterion, acreage could be outside highway ROW and include, for example, an unused wetland bank. Portions of some state ROWs may be ineligible simply because vegetation will not grow in any density, due to soil, hydrology, or local climate limitations (e.g., desert). It is important for the purposes of the project that the acreage not be already enrolled in carbon sequestration efforts.

2. The state DOT selected must be legally able to participate in carbon credit sales.
State laws restricting the state DOT's ability to trade carbon credits should not exist.

3. The state DOT selected should not be from a state with laws already requiring its participation in carbon sequestration.

Qualitative Criteria
Once the essential criteria are used to narrow the candidate field, additional qualitative criteria will be considered. Meeting these criteria is desirable but not essential.

1. The state should have a Climate Action Plan finalized or in revision.

2. The state should belong to an emissions reduction initiative.

There are several cooperative efforts nationwide, such as the Western Climate Initiative and the Regional Greenhouse Gas Initiative, aimed at reducing carbon dioxide emissions. Registries to measure GHG emissions have also been established. The state DOT selected should be a participant in a regional strategy for measuring and/or controlling emissions.

3. The state should have sufficient GIS capability and staff time available to help identify potential project acreage.

Participation in the CSPP would likely require the availability of various subject area experts at the selected DOT (e.g. GIS staff) to assist in the identification of appropriate areas for carbon sequestration activities. The selected DOT should have the staff and funding resources to provide this level of participation. FHWA Headquarters will provide technical assistance.

Considerations for State DOTs
Some questions candidate state DOTs may consider are:
Commitment
If a DOT chooses to package and sell the carbon offsets it develops during the CSPP, a commitment must be made to manage sequestered carbon long-term.

(A) Is state DOT willing to enter into a contract or long-term lease arrangement (5 to 25 years) with a carbon broker or other entity to maintain the carbon sequestered?

(B) If the DOT wishes to sell its carbon credits on the Chicago Climate Exchange, the DOT, state or other entity must commit to emissions reductions. Is state DOT willing to monitor its emissions and, based on its 1999-2000 baseline emissions, agree to a six percent reduction in carbon dioxide by 2010?

Potential Benefits and Costs
The potential benefits and costs of participation in the CSPP will be discussed in some depth during telephone interviews with interested candidates. For some DOTs, the capital cost of planting vegetation in existing ROW or other DOT lands may appear to be a hurdle. However, because appropriately managed biological carbon sequestration projects will reduce water pollution from highway runoff and can be considered as landscaping or scenic beautification, the CSPP may qualify for Transportation Enhancement Funds.

(A) If FHWA provides technical assistance, is the state DOT able to invest in plant materials, planting maintenance, verification of carbon dioxide reductions, staff education, and program administration?

Technical Capacity
Selection for the CSPP likely requires the availability of various subject area experts at the state DOT. GIS and engineering staff would likely be needed to help provide a more precise estimate of available DOT lands. When a state is selected, these staff would help identify specific areas for carbon sequestration, and ensure that the project can be conducted without creating safety hazards or conflicts.

(A) Does the state DOT have the staff and funding resources to provide this level of participation?
Yes [ ]   No [ ]

(B) Does the state DOT have staff with sufficient proficiencies and time to work on this effort?
Yes [ ]   No [ ]

Top State DOT Rankings after Analyzing Selection Criteria

1	Oregon
2	Idaho
3	New Mexico
5	Washington
4	Texas